Mayke da Silva Santos

# Bioecology of Acestrorhynchus britskii (Pisces: Characiformes)

**Mayke da Silva Santos**

# Bioecology of Acestrorhynchus britskii (Pisces: Characiformes)

## Population structure, abundance and reproductive investment in reservoir systems in the São Francisco River basin

**ScienciaScripts**

**Imprint**

Any brand names and product names mentioned in this book are subject to trademark, brand or patent protection and are trademarks or registered trademarks of their respective holders. The use of brand names, product names, common names, trade names, product descriptions etc. even without a particular marking in this work is in no way to be construed to mean that such names may be regarded as unrestricted in respect of trademark and brand protection legislation and could thus be used by anyone.

Cover image: www.ingimage.com

This book is a translation from the original published under ISBN 978-613-9-62454-6.

Publisher:
Sciencia Scripts
is a trademark of
Dodo Books Indian Ocean Ltd. and OmniScriptum S.R.L publishing group

120 High Road, East Finchley, London, N2 9ED, United Kingdom
Str. Armeneasca 28/1, office 1, Chisinau MD-2012, Republic of Moldova, Europe
Printed at: see last page
**ISBN: 978-620-7-72710-0**

# SUMMARY

"With great affection I dedicate it to my loved ones and to all the good-hearted people on earth."


## ACKNOWLEDGMENTS

First of all, to the Great GOD for creating nature and for making my life a blessing every day and for giving me the strength, health and fortitude to carry out this work.

To the State University of Bahia for giving me the opportunity to study a wonderful degree in Biological Sciences.

To CHESF and Laboratòrio Agua e Terra for giving me the permission, the opportunity and the space to carry out this work.

To my dear and special advisor, Eliane Nogueira, for her teachings and suggestions that I will never forget.

To my dear professors Fàtima and Tâmara, who accepted the invitation to sit on the examining board.

To all my family: my mother Dona Saùde, my father José, my brothers Givaldo, Gerlùcia and Janicléia, my nephews Felipe, Heloisa, Joâo Victor, Joabe, Jonatas and Jonadabe, for giving me support and sustenance during this hectic period.

To my fiancée Viviane, for being my inseparable companion at all times, a true blessing in my life.

To all the teachers I've had from primary school up to now, for giving me wise advice and teachings that I'll carry with me for the rest of my life and try to pass on to my future students.

To the teachers I consider to be wonderful people: Nadja, Josilda, Rita and Fernando.

To Professor Marcel for giving me the opportunity to do an internship and scientific initiation in the areas I love so much: zoology and ecology, and for helping me to develop my first research projects.

The 2013.1 and 2012.1 Biological Sciences classes, with whom I had some very good and difficult times during the course.

To my friends:

Clemensou, who always formed our "team of two" with me;

Jonathan, my great companion on the adventures of traveling on the Real Alagoas bus, collections, events and everyday life;

Gilliard and Roberto, my faithful squires who accompanied me during my conducting internships;

Fernando, the coolest vigilante in the world, who always lifted our spirits with his incomparable humor;

Marciany, Jéssica Moreira, Isabella, Leila, Aninha, Jonathans and Vitor, who shared wonderful moments with me in the Agua e Terra laboratory.

And to everyone who contributed directly or indirectly to making it all work.

Thank you very much!

"Do not worry about tomorrow, for tomorrow will take care of itself. Each day's evil is enough.

Matthew 6: 34.

## SUMMARY

*Acestrorhynchus britskii* (dogfish) is a piscivorous species with a preference for lentic environments, endemic to the São Francisco basin. The aim of this study was to assess the population structure of the species in three reservoirs (Sobradinho, Itaparica and Paulo Afonso I, II and III) located on the San Francisco River. The biological material was obtained through six bimonthly collections carried out from July 2015 to May 2016, using a waiting net with meshes of 20 to 140 mm between knots arranged on the surface and on the bottom. We analyzed the Catch Per Unit Effort in numbers (CPUEn), the length structure in 2 cm classes, the sex ratio, the weight-length ratio, the condition factor (K), the length of first maturation ($L_{50}$ ) and the gonadosomatic index (IGS). A total of 206 specimens were caught. The three reservoirs showed similar abundances, with 74 individuals caught in Sobradinho, 72 in Itaparica and 60 in Paulo Afonso I, II and III, with the highest CPUEn in Itaparica (0.0016 ind./$m^2$ ). As for the distribution of the individuals, it was observed that there was a greater catch at the points with lentic flow and the length class with the most representatives was 10-12 cm. The sex ratio did not differ significantly from the expected 1:1, with more males in Sobradinho (0.6F: 1M), and more females in Itaparica (1.5F: 1M) and Paulo Afonso I, II and III (2.5F: IM) during the period analyzed. The species showed negative allometric growth (b<3) in all the reservoirs studied and the condition factor (K) indicated that the population in the Sobradinho reservoir had a better physiological state. The length of first maturation for the Sobradinho, Itaparica and Paulo Afonso I, II and III reservoirs was 12.8 cm, 12.7 cm and 12.1 cm, respectively. The IGS averages revealed that the species shows greater reproductive investment in the Paulo Afonso I, II and III reservoirs (IGS= 4.4), which may be related to the availability of food found there. The results showed that there were no relevant differences between the three populations analyzed, suggesting that the survival conditions for the species are similar in the respective reservoirs, and also that the damming of the river has not yet had a considerable influence on the species, concluding that the availability of resources and reproduction conditions are stable for *A. britskii to* continue maintaining renewable populations in the reservoirs.

**Keywords**: Bioecology. Fisheries biology. Piscivore. Impoundment. São Francisco River.

# 1   INTRODUCTION

The main hydrographic basins in Brazil are being regulated by the construction of reservoirs, which alone or in cascades cause important qualitative and quantitative impacts on the main ecosystems of inland waters (BARBOSA, 2010). Large reservoirs are used for various purposes: hydroelectricity, water reserves for irrigation, drinking water reserves, biomass production (fish farming and intensive fishing), transportation (waterways), recreation and tourism (PAIVA *et al.*, 1994).

The construction of dams is capable of producing major changes in the biota of inland waters in Brazil. In addition to damming, many reservoirs have been repopulated with exotic species, making the food web, community composition and commercial exploitation extremely complex (AGOSTINHO, PELICICE; GOMES, 2008).

Impoundments fragment the habitat, causing genetic isolation of populations after construction. This loss of genetic variability reduces reproductive success, survival and genetic diversity, decreasing the ability of populations to respond to environmental changes, thus altering the composition and structure of populations and mainly affecting fish assemblages (SATO *et al.*, 2005, TOS *et al.*, 2014).

The Sâo Francisco River has five hydroelectric power stations that dam its main course: Três Marias, Sobradinho, Itaparica, Paulo Afonso Complex and Xingó (ANA, 2016). The ichthyofauna of this basin is very threatened in some regions, mainly in the stretch downstream of the Sobradinho dam to the Atlantic Ocean, upstream of the Três Marias dam and in the tributary rivers Paraopeba and das Velhas, mainly due to riparian deforestation, the construction of dams, industrial and domestic pollution, overfishing and the destruction of floodplains and marginal lakes (SATO; GODINHO, 1999).

The notorious ichthyic diversity of the Sâo Francisco basin presents the order Characiformes as the second largest in terms of number of species (BARBOSA; SOARES, 2009), with Acestrorhynchidae as one of the most prominent families with two species belonging to the genus *Acestrorhynchus,* which is also distributed in other South American river basins (REIS, KULLANDER; FERRARIS, 2003; BERRA, 2007). The species in this genus are ichthyophagous (MENEZES, 1992) and are important for the ecosystem, forming part of the diet of large, commercially valuable piscivores and also controlling forage species (PERET, 2004).

*Acestrorhynchus britskii* (dogfish) is a species endemic to the Sâo Francisco basin. They vary in length from 10 to 19 cm and can reach 55 g in adulthood. It has a piscivorous feeding habit, preying on shoals and has a preference for lentic environments (MENEZES, 2003; ROCHA *et al.,* 2015). It has no economic importance in the region, although it serves as a protein supplement for riverside populations.

Population structure refers to the density and distribution of individuals in each age class (RICKLEFS, 2010). Studies on the population structure of fish are important as they can provide information on the ecology, habits and functions of the species in its natural environment, generating data that allows for efficient management within the ecosystem (BENEDITO- CECILIO; AGOSTINHO, 1997).

Information on the population structure of an endemic species such as *Acestrorhynchus britskii* (dogfish) is necessary in order to understand the ecological dynamics of fish communities in reservoirs, and can thus support appropriate management measures for the maintenance of aquatic biota, as the loss of important components such as this can affect the functioning of the ecosystem in the future. It also provides information that will allow us to know whether individuals are managing to maintain renewable populations in reservoirs, as well as offering subsidies for conservation programs.

# 2  OBJECTIVES

## 2.1  General

To compare the population structure of *Acestrorhynchus britskii* (Menezes) in the Sobradinho, Itaparica and Paulo Afonso I, II, III reservoirs located in the São Francisco River basin, Brazil.

## 2.2  Specifics

• Estimate the abundance (CPUEn) of the species in the reservoirs;

• Analyze the length structure of the population, sex ratio, weight-length relationship and condition factor of the species in the reservoirs;

• Estimate the average length at first maturity;

• To see if there is a difference in reproductive investment between the populations.

# 3 THEORETICAL BASIS

## 3.1 Neotropical Ichthyofauna

Striped finned fish (Osteichthyes: Actinopterygii) comprise the most diverse and numerous group of vertebrates, including approximately 33,000 valid species (ESCHMEYER; FONG, 2016) distributed in 453 families (NELSON, 2006). They have adaptations that have allowed them to colonize different aquatic habitats and these adaptations are reflected in the anatomical, physiological, behavioural and ecological variations presented by the group (HELFMAN *et al.*, 2009).

Among freshwater fish, the Neotropical ichthyofauna is highly diverse, mainly due to the presence of large hydrographic systems, with considerable ichthyofaunal differentiation between them. However, much of this diversity is still unknown (AGOSTINHO *et al.*, 2007), and it is estimated that 20% is critically endangered due to pollution, the introduction of exotic species and habitat loss (HILSDORF; PETRERE, 2002).

Ichthyic diversity in Brazil is remarkable, especially in the São Francisco River, whose basin drains areas in the states of Minas Gerais, Bahia, Pernambuco, Alagoas, Sergipe, Goiàs and the Federal District, cutting across three biomes: Cerrado, Caatinga and Atlantic Forest. For the area, 244 valid fish species are listed, distributed in 35 families and 10 orders. Of this total, there are 214 native species, 24 introduced species and six estuarine species. The orders Siluriformes and Characiformes are the most representative in terms of number of species with 86 and 85 respectively, followed by Cyprinidontiformes with 43, Perciformes with 18, Gymnotiformes with 4, Cupleiformes with 3, Cypriformes with 2, and the others Pleuronectiformes, Symbranchiformes and Lepidosireniformes with one species each (BARBOSA; SOARES, 2009).

With an extension of approximately 2,900 km (GODINHO; GODINHO, 2003), the São Francisco basin has a highly diverse biota, with very relevant endemism observed among fish, in all there are 76 endemic species, representing 36.6% of its ichthyofauna, with emphasis on members of the Rivulidae family known as annual fish, which represent 62.5% of the endemic species of the Sâo Francisco (ALVES; POMPEU, 2005).

Several genera of fish found in the Sâo Francisco basin are common to the Amazon, Parnaiba and Paranâ-Uruguay basins. However, there is greater ichthyofaunal similarity between the Sâo Francisco River and the rivers of the Pardo/Mogi-Guaçu system (Paranâ-Uruguay basin), than between the latter and the Parnaiba River (Maranhâo/Piaui) (BARBOSA, 1986). On the other hand, there are already established populations of introduced species, such as *Cichla* sp. (tucunaré), *Plagioscion squamosissimus* (corvina), *Cyprinus carpio* (carpa), *Clarias gariepinnus* (catfish), *Colossoma macropomum* (Tambaqui), *Oreochromis* sp. and *Tilapia* sp. (tilapias), among others (SATO; GODINHO, 1999).

The damming of the river and overfishing have considerably affected the fish populations of the Sâo Francisco, especially migratory species, while other factors such as riparian deforestation, industrial and domestic pollution, mining and the destruction of floodplains are also pointed out as real threats to fish in the basin (SATO; GODINHO, 1999; MELO *et al.*, 2011).

The ichthyofauna of this basin has been studied since the first scientific expeditions to Brazil, such as Lutken (1875). Since then, the works of Travassos (1960), Britski, Sato; Rosa (1988), Godinho (2009) and Barbosa; Soares (2009) have been of great relevance to the basin, presenting lists of species and identification keys. And although there has been a long period of studies on the ichthyofauna of the Sâo Francisco River, basic studies on fish biology are still incipient in this basin as it is severely impacted and species become extinct (BRITSKI, SATO; ROSA, 1988).

## 3.2 Order Characiformes

The Ostariophysi superorder is a vast lineage of more than 8,000 species that dominates freshwater fish faunas around the world, but also includes more than 100 marine species. The lineage comprises five orders: Gonorynchiformes, Gymnotiformes, Characiformes, Cypriniformes, and Siluriformes (HASTINGS, WALKER; GALLAND, 2014).

The Characiformes make up an extremely diverse lineage of freshwater fish with 23 families, 270 genera and approximately 2,100 species, mostly distributed in South and Central America and southern North America, with less representation in Africa where it also occurs naturally (VIEIRA *et al.*, 2015 ; ESCHMEYER; FONG, 2016).

The species of this order are characterized by having a body covered in scales except for the head, a premaxillary usually fixed to the skull and not protruding, multicuspidate maxillary teeth, three to five gill rays, fins without spines, a pelvic fin located in an abdominal position, thus being a ventral fin, and almost always the presence of an adipose fin located between the dorsal and caudal fins (BRITSKI, SATO; ROSA, 1988; HASTINGS, WALKER; GALLAND, 2014).

The Characiformes have a variety of ecological specializations, evidenced by the enormous variety of shapes, sizes and feeding habits. They are of notable ecological importance due to their abundance and diversity, and many are commercially important as food and ornamental fish (LOWE-MCCONNELL, 1999; MOTA; PRIOLI; PRIOLI, 2014).

In the Sâo Francisco basin, 85 species and 45 genera of Characiformes are listed, distributed in 12 families: Acestrorhynchidae, Anostomidae, Bryconidae, Characidae, Curimatidae, Crenuchidae, Erythrinidae, Hemiodontidae, Iguanodectidae, Parodontidae, Prochilodontidae and Triportheidae. The Characidae family stands out with 46 species and 31 genera and the *Leporinus* genus with 10 species, which belongs to the Anostomidae family (BARBOSA; SOARES, 2009; ESCHMEYER; FONG, 2016).

### 3.3 Taxonomic and ecological description of *Acestrorhynchus britskii* (Menezes)

The Acestrorhynchidae family currently has three subfamilies: Acestrorhynchinae, Roestinae and Heterocharacinae, together totaling 26 species distributed in 7 genera (ESCHMEYER; FONG, 2016). Acestrorhynchinae is the most specious, with *Acestrorhynchus* as its only genus, represented by 14 valid species occurring in South America (PRETTI *et al.,* 2009), with the greatest specific diversity occurring in the Amazon and Orinoco basins (11 species), as well as the drainages of the San Francisco River (two species), Paranà, Paraguay and La Plata (one species) (REIS, KULLANDER; FERRARIS, 2003 ; BERRA, 2007).

The genus *Acestrorhynchus* belongs to a homogeneous tribe, and most species can be identified by various external characteristics, including snout length, coloration and maximum size. All forms are ichthyophagous and require oxygenated, clear water with a temperature of between 2328°C (MENEZES, 1992). Although they have no commercial value, *Acestrorhynchus* species are part of the diet of valuable fishing species, as well as being controllers of forage fish populations (PERET, 2004).

Of the two species of the genus *Acestrorhynchus* occurring in the São Francisco basin, *Acestrorhynchus britskii* (dogfish) (Figure 1) deserves to be highlighted, as it is endemic to the basin (MENEZES, 2003), and also because it is one of the most representative species numerically in the Sobradinho reservoir (FADURPE, 2004).

**Figure 1. *Acestrorhynchus britskii* in left lateral position. 1- Gray longitudinal stripe, 2- Black spot on caudal peduncle.**

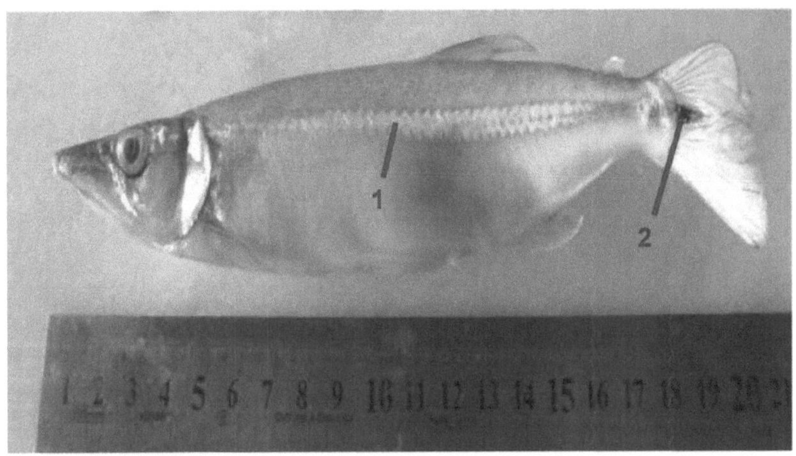


**Photo: SANTOS (2016).**

It is characterized by having an elongated and compressed body, a forked caudal fin, a large and terminal mouth with a premaxilla with canine teeth separated from each other by a series of conical teeth, a maxilla with conical teeth along its entire edge and a canine tooth at the proximal end. A gray longitudinal stripe and a black spot at the end of the caudal peduncle which can extend to the tip of the median caudal rays (BRITSKI, SATO; ROSA, 1988).

This species prefers lentic environments, such as lakes and lagoons, and can also be found in some stretches of rivers. It has a piscivorous feeding habit and preys in groups (SATO; GODINHO, 1999; MENEZES, 2003). **The** study by Barros; Santos (1996) on the reproductive biology of *Acestrorhynchus britskii* found that the species has low fecundity and spawns in installments, as well as having a higher sex ratio of females than males and reproductive peaks during the period of highest temperature and hydrological level in the Três Marias reservoir on the upper São Francisco.

## 3.4 Fish population structure

A population consists of individuals of a species in a given geographical area (SARMIENTO, 2000). However, it represents more than a collection of individuals, it has integrity as an organizational unit in ecology because individuals come together to reproduce, thus mixing the population's gene *pool* and ensuring its continuity over time (WOOTTON, 1990; RICKLEFS, 2010).

Individuals in a population can be distributed according to three major patterns: casual, uniform and aggregated. Aggregation in varying degrees is the most common pattern, almost the rule when considering individuals (ODUM, 2007). Aggregation in fish has functions such as reducing predator success, increasing foraging success and synchronizing reproductive behaviour (HELFMAN *et al.*, 2009).

In fish populations, knowledge of quantitative aspects such as: abundance, weight-length ratio, condition factor, sex ratio, growth, average length at first maturity, recruitment and mortality, provide basic information for the study of fisheries biology, important for rational fisheries management in an environment (LE CREN, 1951; NIKOLSKII, 1969 ; WOOTTON, 1990; LIZAMA ; AMBRÓSIO, 1999).

Some studies in South America have analyzed fish populations and the impact of dam construction on them (GODINHO; GODINHO, 1994; PETRERE, 1996; DE MÉRONA; ALBERT 1999; SATO *et al.* 2005; ARANTES *et al.* 2010). In the semi-arid region of Brazil, studies such as Nascimento; Gurgel (2000); Raposo; Gurgel (2001); Gurgel (2004); Santos; Novaes (2008); Chaves *et al. (*2009) and Montenegro *et al.* (2011) analyzed the population structure of teleost species in rivers in the states of Paraiba and Rio Grande do Norte. However, in important river basins in the semi-arid region, such as the Sâo Francisco and the Parnaiba, which are used by hydroelectric dams and cut through important states in the Northeast, studies analyzing the population structure and age of fish populations are scarce.

# 4 MATERIAL AND METHODS

## 4.1 Study area

The São Francisco basin is located between parallels 7° 00' and 21° 00' S and 36° 15' and 47° 39 W (Figure 2), which gives it very varied climatic characteristics, with rainfall ranging from 350 to 1,900 mm in normal years. Average annual temperatures range from 18° to 27° C and evaporation is relatively high, ranging from 2,300 to 3,000 mm per year. The average relative humidity is between 60 and 80% and luminosity is high, from 2,400 to 3,300 hours a year [1] (CODEVASF, 1991).

**Figure 2 - Map of the São Francisco river basin.**

**Source: São Francisco Hydrographic Basin Committee (CBHSF), 2015.**

The basin is divided into four physiographic regions: upper, middle, sub-middle and

15

lower. The upper course is characterized by fast, cold and oxygenated waters; the middle course, being a plateau river, is slower and subject to major floods; the middle course is practically dammed and the lower course, being a plateau stretch, is slower and under marine influence (SATO; GODINHO, 1999).

This study analyzed three cascade reservoirs located in the São Francisco basin: the Sobradinho reservoir (BA, PE), the Itaparica reservoir (PE) and the Paulo Afonso I, II and III reservoirs (BA) (Figure 3).

The Sobradinho hydroelectric dam is located on the middle São Francisco River, 748 km from its mouth. In addition to generating electricity, it is also the main source of regularizing the region's water resources. It is about 320 km long, with a water surface of 4,214 km$^2$ and a storage capacity of 34.1 billion cubic meters at its nominal level of 392.50 m, making it the largest artificial lake in the world (CHESF, 2017a).

The Itaparica hydroelectric dam, which was renamed Luiz Gonzaga in honor of the "king of the northeastern bayou" of the same name, is located in the São Francisco sub-medium 50 km upstream from the Paulo Afonso Hydroelectric Complex. In addition to generating electricity, it also regulates the daily and weekly inflows from those plants. The top of the dam is at 308.10 m with a crest width of 10.00 m (CHESF, 2017b).

The Paulo Afonso I, Paulo Afonso II and Paulo Afonso III power plants are located in the same impoundment in the São Francisco sub-medium, consisting of a reinforced concrete gravity dam with a maximum height of 20 m and a total crest length of 4,707 m (CHESF, 2017c).

**Figure 3 - Map of the location of the reservoirs studied.**

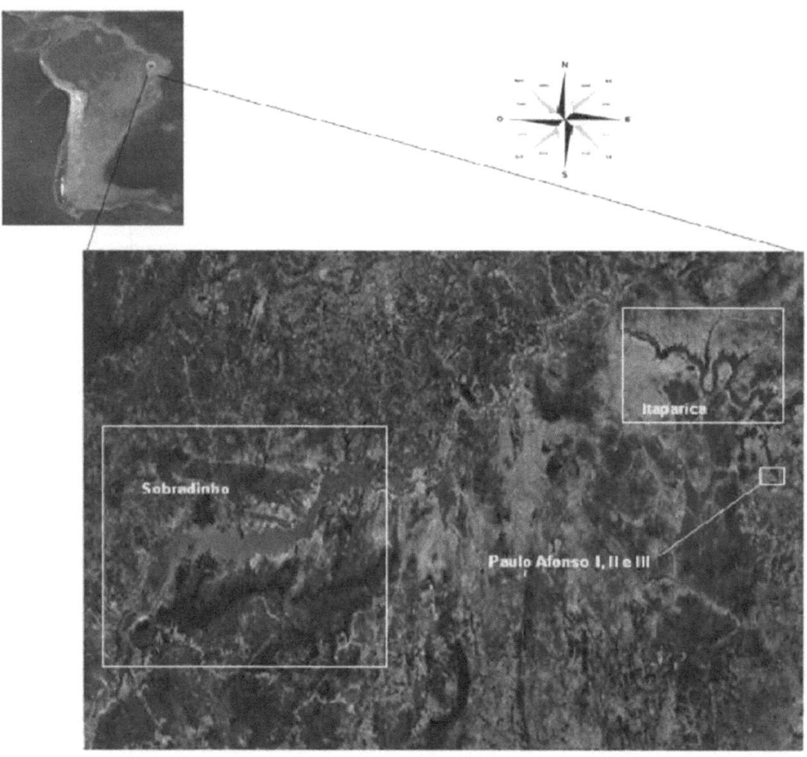

**Source: Modified Google Earth, 2017.**

Collections took place at 21 sampling points in eight municipalities in the states of Bahia and Pernambuco (Table 1), with nine points in the Sobradinho reservoir, eight in the Itaparica reservoir and four in the Paulo Afonso I, II and III reservoirs (Figure 4).

**Table 1. Collection points and their respective municipalities, states, coordinates and type of flow.**

| SOBRADINHO RESERVOIR | | | |
|---|---|---|---|
| POINT | MUNICIPALITY/STATE | COORDINATES (UTM) | FLOW |

| P1 | Santa Maria da Boa Vista-PE | 24L 9025625 | 0409924- | Lòtico |
|---|---|---|---|---|
| P2 | Belém do Sao Francisco-PE | 24L 9027878 | 0508602- | Transition |
| P3 | Santa Maria da Boa Vista-PE | 24L 9026384 | 0411081- | Lòtico |
| P4 | Sobradinho-BA | 24L 8962375 | 0301052- | Lentic |
| P5 | Remanso-BA | 23L 8930791 | 0821626- | Lentic |
| P6 | Remanso-BA | 23L 8930710 | 0826752- | Lentic |
| P7 | Belém do Sao Francisco-PE | 24L 9026647 | 0505882- | Transition |
| P8 | Xique-xique-BA | 23L 8800705 | 0746933- | Lentic |
| P9 | Barra-BA | 23L 8772984 | 0703813- | Lòtico |
| **ITAPARICA RESERVOIR** | | | | |
| P1 | Petrolândia-PE | 24L 9005981 | 0583375 | Lentic |
| P2 | Petrolândia-PE | 24L 8990334 | 0577877 | Lentic |
| P3 | Petrolândia-PE | 24L 8989444 | 0573480 | Lèntic |

| P4 | Belém de Sao Francisco-PE | 24L 9027488 | 0507076 | Transition |
|---|---|---|---|---|
| P5 | Petrolândia-PE | 24L 9005886 | 0586314 | Lentic |
| P6 | Petrolândo-PE | 24L 9002744 | 0580742 | Lentic |
| P7 | Belém de Sao Francisco-PE | 24L 9027921 | 0508776 | Transition |
| P8 | Belém de Sao Francisco-PE | 24L 9027981 | 0502816 | Transition |
| **PAULO AFONSO RESERVOIR I, II AND III** | | | | |
| P1 | Paulo Afonso-BA | 24L 8964348 | 0587052 | Lentic |
| P2 | Paulo Afonso-BA | 24L 8963252 | 0587469 | Lentic |
| P3 | Paulo Afonso-BA | 24L 8956948 | 0580506 | Lentic |
| P4 | Paulo Afonso-BA | 24L 8958941 | 0584870 | Lentic |

**Figure 4: Sampling stations at the Sobradinho reservoir (A), Itaparica (B) and Paulo Afonso I, II and III.**

**(A)**

Photos: **NÓBREGA (2016).**

(B)

**(C)**

**Photos: NÓBREGA (2016).**

## 4.2 Collection and fixing

To analyze the population structure of *Acestrorhynchus britskii,* bimonthly collections were carried out from July 2015 to May 2016, totaling six collections in the Sobradinho, Itaparica and Paulo Afonso I, II and III reservoirs. The catches were made using waiting nets with meshes of 20, 30, 40, 50, 60, 70, 80, 100, 120 and 140 mm between knots, arranged in surface and bottom batteries which were exposed for 12 hours at each sampling point, being set at dusk and removed at dawn. The specimens collected were identified according to Britski, Sato; Rosa, (1988) and then fixed in 10% formaldehyde solution and transported to the Ichthyofauna Laboratory of the company Laboratòrio Agua e Terra LTDA located in Paulo Afonso-BA. After analysis, the individuals were deposited in the São Francisco River Reference Collection (CRSF) located at UNEB *campus* VIII - Paulo Afonso-BA.

## 4.3 Laboratory tests

The specimens collected were measured (mm) and weighed (g) and then dissected for sex identification with a median incision starting at the urogenital opening and going up to the isthmus. The gonads, on both sides, were removed, then weighed on a precision scale and classified macroscopically as to their maturation stage, following a scale adapted from Vazzoler (1996), considering four stages:

> **Stage A or immature:** The ovaries and testicles are translucent, very small and placed very close to the vertebral column.

> **Stage B or maturing:** The ovaries occupy about 1/3 to 2/3 of the abdominal cavity, with an intense capillary network, and there are opaque granules of varying sizes, the oocytes. The testicles, like the ovaries, are developed and lobulated.

> **Stage C or mature:** Ovaries and testicles are enlarged, almost completely occupying the abdominal cavity, and mature oocytes are seen as opaque or translucent spherical granules.

> **Stage D or spawned:** The ovaries and testicles have an emorrhagic appearance, are completely fluid and occupy less than 1/3 of the abdominal cavity.

## 4.4 Data analysis and statistics

Abundance was estimated using the numerical data of the species caught in the reservoirs and using the Catch Per Unit Effort in number (CPUEn): CPUE = C/f; C= catch in number; f= effort $m^2$ x h, where $m^2$ = 608 m2 of net and h= 12 hours. To analyze the population structure in terms of length, the individuals were separated into 2 cm length classes. To analyze the sex ratio, adult individuals were used and the absolute frequency was obtained by counting the number of females and males and calculating the relative frequency (%) of each sex.

To detect the species' growth strategies, we used the length-weight relationship determined by the following equation: Wt = $a$ Lp$^b$ ; Wt is the total weight (g); Lp is the standard length (cm); $a$ *is* the intercept and $b$ is the straight line. The condition factor was estimated using the equation K = (Wt/Lpb) x 100; where K is the condition factor, Wt is the total body weight; Lp is the standard length and b is the angular coefficient (LE CREN, 1951).

To estimate the length of first maturation, grouped sex was used, the fish were separated according to maturation stage into "juveniles" (immature stage) and "adults" (other stages), and the length of first maturation was calculated using the generalized linear model (GLM) by fitting a logistic regression equation, with standard length as the independent variable. Pmaduro = exp (a + b + L) x [1 + exp (a + b + L)] - 1, where Pmaduro = estimated proportion of mature individuals in relation to standard length (L). The coefficients were estimated by least squares redistribution analysis.

The reproductive investment of the species in the reservoirs was measured using the gonadosomatic index (GSI), using only mature females, which expresses the percentage of the gonads in the total weight of the individuals, and is given by the formula: GSI = (Wg/Wt) x 100; Wg is the weight of the gonads and Wt is the weight of the body (WOOTTON, 1990).

Student's t-test was used to compare the mean abundance (CPUEn) of the populations, the mean length of first maturation and the general value of IGS in the three reservoirs. The Chi-squared test ($\chi^2$) was used to test for differences in sex ratio.

All the analyses were carried out using Microsoft Office Excel 2010 software with a significance level of p<0.05 and details can be found in Gotelli; Ellison (2011).

## 4.5 Forms of general graphic presentation

In terms of graphic presentation, this monograph follows the recommendations of NBR 14724 (ABNT, 2011).

# 5    Results and discussion

Of the total of 206 specimens, 74 were caught in the Sobradinho reservoir, with CPUEn values ranging from 0.0005 (January/2016) to 0.0031 (September 2015)(0.0013 ± 0.0009) individuals $m^2$ .h, 72 in the Itaparica reservoir, with CPUEn values between 0.0001 (March/2016) and 0.0038 (July/2015) (0.0016 ± 0.0014) individuals m2.h, and 60 in the Paulo Afonso reservoir (PA) I, II and III with CPUEn values between 0.0006 (November/2015) and 0.0021 (September/2015) (0.0013 ± 0.0006) individuals m2.h (Table 1)(Figure 5).

**Table 1. Number of specimens of *Acestrorhynchus britskii* caught and average CPUEn in the Sobradinho, Itaparica and PA I, II and III reservoirs.**

|  | sobradinho | Itaparica | PA I, II and III | Total |
|---|---|---|---|---|
| **No. of individuals captured** | 74 | 72 | 60 | 206 |
| **Average CPUEn** | 0.0013 ind./$m^2$.h | 0.0016 ind./m2.h | 0.0013 ind./m2.h | |

Figura 5. **Graph of CPUEn values by month of collection in the Sobradinho, Itaparica and PA I, II and III reservoirs.**

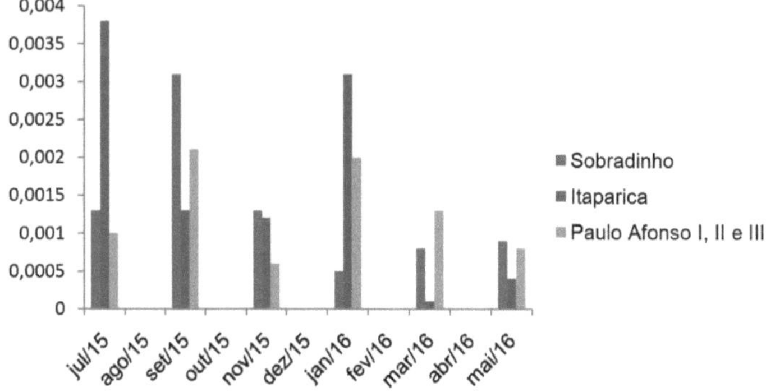


The t-test showed that the abundance of the species was statistically equal between the reservoirs with t= 0, 6526, between Sobradinho and Itaparica; t= 0, 9716, between Sobradinho and PA I, II and III; and t= 0, 6117, between Itaparica and PA I, II and III, with a significance level of p> 0.05. The equality of the abundance values indicates that the survival conditions for the species are similar in the reservoirs analyzed.

The largest catches were observed in July (Itaparica) and September (Sobradinho and PA I, II and III) 2015, months in which the Sâo Francisco is at its lowest ebb, and also in January 2016 (Itaparica and PA I, II and III), when the river is at its lowest ebb, so it is not possible to determine an ideal period for catching the species.

Analysis of the spatial distribution of the species in the reservoirs revealed that in Sobradinho there was a higher catch at points 4, 5, 6 and 7 and lower catch rates at points 8 and 9, in Itaparica there was a higher catch at point 5 and a lower catch at point 1 and in PA I, II and III there was a higher catch at points 3 and 4 and a lower catch at point 2 (Figure 6).

Figura 6. **Graph showing CPUEn per collection point in the Sobradinho (a), Itaparica (b) and PA I, II and III (c) reservoirs.**

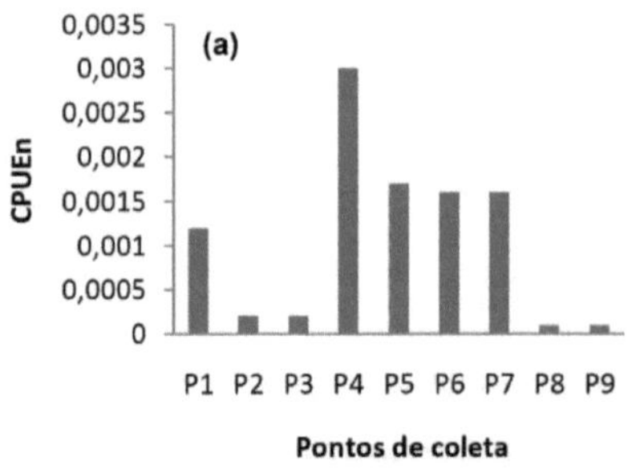


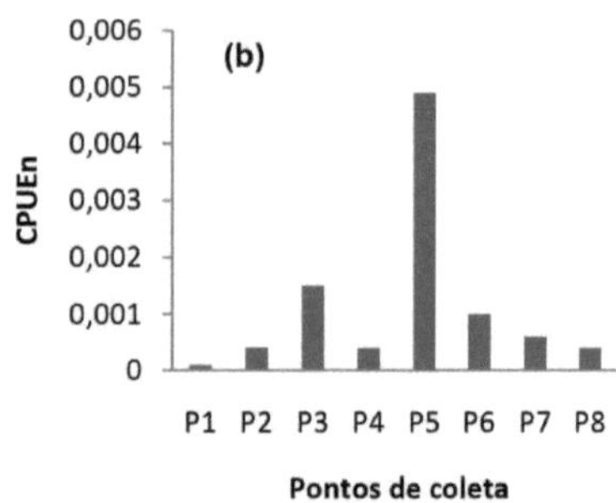

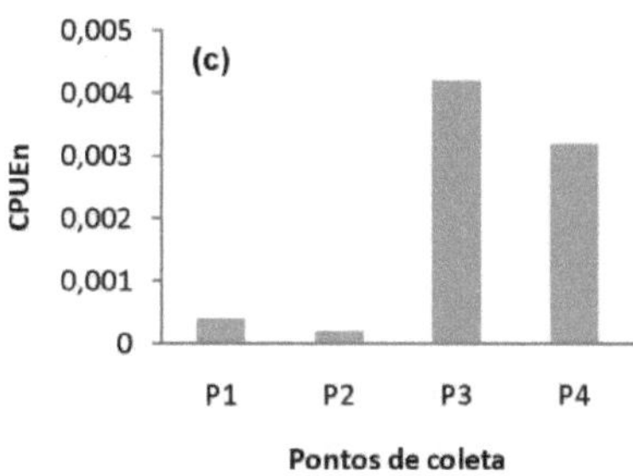


In both reservoirs, distribution was heterogeneous and it was observed that the points with the highest catch rates were those located in lentic and transitional flow regions, corroborating information in the literature describing *A. britskii* as a species that prefers lake and pond environments (SATO; GODINHO, 1999; MENEZES, 2003).

The standard length of the individuals collected in the Sobradinho reservoir varied between 108 and 170 mm (130.8 ± 9.1 mm) and the weight varied between 12 and 57 g (25.7 ± 15.5 g). In Itaparica, the standard length varied between 102 and 175 mm (131.2 ± 21.2 mm) and the weight varied between 13 and 51 g (23.2 ± 14.8 g). In the PA I, II and III reservoirs, the standard length ranged from 103 to 180 mm (125.8 ± 2.8 mm) and the weight ranged from 12 to 62 g (21.6 ± 2.1 g).

The distribution of individuals by length class was homogeneous between the reservoirs. The highest occurrence was in the 10-12 cm class, followed by 13-15 cm and 16-18 cm (Figure 7).

The composition of a population in length classes is a characteristic that responds to the environment and can vary from year to year, depending also on its fecundity (NIKOLSKII, 1969) and can provide subsidies for the study of the determination of the balance of this population, involving estimates of the rate of mortality, reproduction, recruitment and growth.

Figura 7. **Graph showing the distribution of *Acestrorhynchus britskii* individuals by length class in the Sobradinho, Itaparica and PA I, II, III reservoirs.**

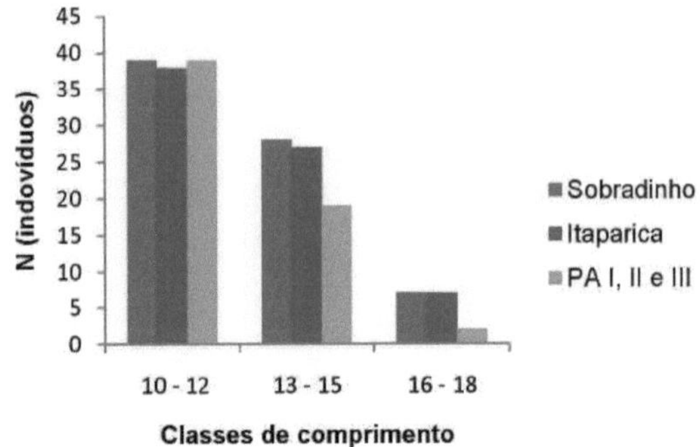


As in the study by Rocha *et al.* (2015), specimens over 16 cm were also observed, which was the maximum length of the species cited in the literature until this study was carried out.

Among the individuals caught, the females were larger than the males in terms of length and weight (Table 2), a sexual dimorphism already observed in neotropical Characiformes (HOJO *et al.*, 2004; THOMÉ *et al.*, 2005; ALVARENGA et al., 2006; GONÇALVES *et al.*, 2006; CARVALHO et *al.*, 2009). According to Vazzoler (1996), females are in the largest length classes because they have a higher growth rate than males, so they reach a larger size than males at the same age.

**Table 2. Biometry of females and males of *Acestrorhynchus britskii* in the Sobradinho, Itaparica and PA I, II and III reservoirs. CP=Standard length; PT=Total weight; N=number of individuals.**

| | CP (mm) | | PT (g) | | N (individuals) | |
|---|---|---|---|---|---|---|
| Reservoirs | Females | Males | Females | Males | Females | Males |
| Sobradinho | 143 ± 14,6 | 123 ± 10,9 | 33,3 ± 10,1 | 20,8 ± 8,09 | 29 | 45 |
| Itaparica | 135,8 ± 17,3 | ±124,5 13,3 | ± 26,2 ± 10,4 | 18,8 ± 6,5 | 44 | 28 |
| PA I, II and III | 130,3 ± 15,0 | 114,5 ± 5,9 | 23,7 ± 6,9 | 14,8 ± 2,5 | 43 | 17 |

In the Sobradinho reservoir, 45 males and 29 females were caught with a ratio of 0.6F: 1M ($\chi^2$ = 0.0307). In Itaparica, 28 males and 44 females were caught with a ratio of 1.5F: 1M ($\chi2$ = 0.0250). And in PA I, II and III, 17 males and 43 females were caught with a ratio of 2.5F: 1M ($\chi2$ = 1.4248) (Table 3).

Various factors can play a role in determining the sex ratio in fish. Mortality, growth and behavior are examples of factors that act differently on the sexes, which can alter the sex ratio at various stages of development (NASCIMENTO, YAMAMOTO; CHELLAPPA, 2012).

In most studies, a 1:1 ratio is observed, but a factor such as the population's food supply can be considered an important factor. In oligotrophic aquatic environments, males predominate and females predominate when food is abundant (NIKOLSKII, 1969).

The results presented suggest that during the sampling period, the PA I, II and III reservoirs showed greater availability of food for the species, since the proportion of females was higher in practically all the collection months, with the exception of March and May 2016, when the proportion was the same for both sexes. In the study by Barros; Santos (1996) carried out in the upper Sâo Francisco, more females than males of *A. britskii were* also observed during the period analyzed. In the other two reservoirs, Sobradinho proved to be oligotrophic during this period, a fact already observed by Lima; Severi (2014), with a superiority of males, while Itaparica showed a balance approaching the expected ratio of 1:1.

**Table 3. Sex ratio (SP) by month of collection of *Acestrorhynchus britskii* in the Sobradinho, Itaparica and Paulo Afonso Complex I, II and III reservoirs. F=female, M=male.**

| | Sobradinho | | | | Itaparica | | | | PA I, II and III | | | |
|---|---|---|---|---|---|---|---|---|---|---|---|---|
| Month | F | M | Total | PS | F | M | Total | PS | F | M | Total | PS |
| Jul/15 | 4 | 6 | 10 | 0,6:1 | 19 | 9 | 28 | 2:1 | 6 | 2 | 8 | 3: 1 |

| Sep/15 | 9 | 19 | 28 | 0,6:1 | 5 | 5 | 10 | 1:1 | 16 | 0 | 16 | 0 |
|---|---|---|---|---|---|---|---|---|---|---|---|---|
| Nov/15 | 5 | 5 | 10 | 1:1 | 7 | 2 | 9 | 3:1 | 4 | 1 | 5 | 4: 1 |
| Jan/16 | 1 | 3 | 4 | 0,3:1 | 8 | 12 | 20 | 0,6:1 | 10 | 5 | 15 | 2: 1 |
| Mar/16 | 7 | 8 | 15 | 0,8:1 | 1 | 0 | 1 | 0 | 4 | 6 | 10 | 0,6: 1 |
| May/16 | 3 | 4 | 7 | 0,7:1 | 2 | 1 | 3 | 2:1 | 3 | 3 | 6 | 1: 1 |

The weight-length relationship equation for the *Acestrorhynchus britskii* population in the Sobradinho reservoir was Wt= -55.7183 Lp$^{1,34721}$ and correlation coefficient r= 0.915932. In the Itaparica reservoir, the weight-length equation was Wt= -48.2925 Lp$^{1,614696}$ and r= 0.93802. In PA I, II and III, the equation was Wt= -51.5134Lp$^{1,516081}$ and r= 0.936141 (Figure 2).

According to the scale proposed by Orsi, Shibatta and Silva-Souza (2002), when the value of b (angular coefficient) = 3 growth is isometric, b < 3 is negative allometric and b > 3 is positive allometric. In all three reservoirs, the populations showed a value of b < 3, indicating that the species has negative allometric growth, where there is a smaller increase in weight than in length.

Figura 8. **Graphs of the weight-length relationship of *Acestrorhynchus britskii* in the Sobradinho (a), Itaparica (b) and PA I, II and III (c) reservoirs.**

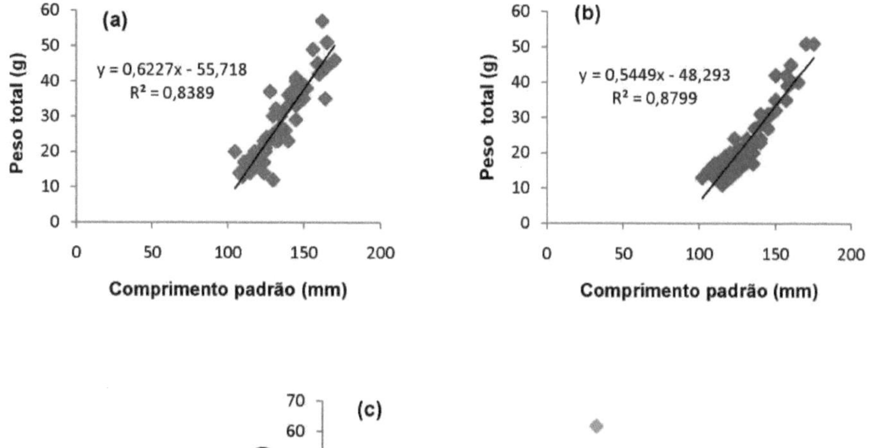



The type of growth is related to different environmental conditions, reflecting the species' well-being and different ways of using resources and allocating energy, with the aim of obtaining better conditions for remaining in the environment (SILVA et al., 2010). Biometric variables may be influenced by various factors such as population density, food availability and abiotic factors characteristic of each environment which, when interacting with each other, may affect the estimated values of the relationship (SOUZA et al., 2000).

In the work carried out by Santos, Vilhena-Picanço ; Sà-Oliveira (2007), the weight-length relationship of *Acestrorhynchus lacustris* was analyzed in the Curiaù River - AP, where negative allometric growth was also observed, indicating that this type of growth strategy is typical of the genus.

In the case of *A. britskii,* this type of growth may be related to its reproductive and feeding strategy. Barros; Santos (1996) report that the species has low fecundity, characteristics that do not lead to large increases in weight during the period of gonad maturation. According to Begon, Townsend and Harper (2007), weight gain exceeding length gain is not advantageous for active predators, so for *A. britskii*, which is a piscivore that preys on shoals (MENEZES, 2003), negative allometric growth is probably an adaptation due to its feeding habits and the fact that it is a predator.

The average condition factor values were K= 3.689112 for the Sobradinho population, K= 0.97680 for the Itaparica population and K= 1.394633 for the PA I, II and III populations (Figure 9).

The condition factor can be defined as the state of well-being of the fish, i.e. how the animal takes advantage of the available resources at a given time of year. It is often used as an indicator of the spawning period, since in this period feeding intensity ceases and the condition factor shows low values (BARBIERI, HARTZ; VERANI, 1996).

**Figura 9. Graph showing the condition factor (K) of *Acestrorhynchus britskii* in the Sobradinho, Itaparica and PA I, II and III reservoirs.**

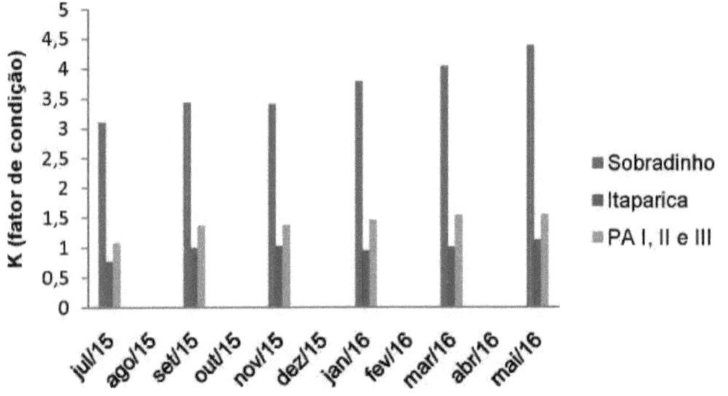


The highest condition factor values were observed in the individuals from Sobradinho, indicating that during this period they made better use of the resources available in the environment than the individuals from the other two reservoirs. The differences observed may have been due to various factors such as availability and competition for resources, interactions with other species and factors related to the environment such as water temperature, conductivity, dissolved oxygen, habitat and seasonality (GUBIANI; HORLANDO, 2014).

The length of first maturity ($L_{50}$ ) for grouped sexes in the Sobradinho reservoir was estimated at 128 mm, the largest immature individual caught was 142 mm and the smallest mature was 110 mm. In Itaparica, first maturity was estimated at 127.5 mm, the largest immature individual was 136 mm and the smallest mature was 102 mm. And in PA I, II and III, first maturity was estimated at 121 mm, with the largest immature individual measuring 120 mm and the smallest mature measuring 103 mm (Figure 10).

The comparison between the three populations showed that there was no significant difference between the lengths of first maturity (t=0.992949, p>0.05, between Sobradinho and Itaparica, t= 0.059876, p>0.05, between Sobradinho and PA I, II and III, and t=0.067722, p>0.05, between Itaparica and PA I, II and III).

In all the reservoirs analyzed, it was observed that approximately 80% of the individuals were above the size of first sexual maturity (L50), indicating the reproductive activity of the species during the period studied, which may also reflect an important adaptive character, since, in unstable environments such as reservoirs, the species suffer greater environmental pressure due to greater variations in water level than in natural environments (SUZUKI *et al.*, 2009),

Figura 10. **Length of first maturation of *Acestrorhynchus britskii* for grouped sexes, in the Sobradinho (a), Itaparica (b) and PA I, II and III (c) reservoirs.**

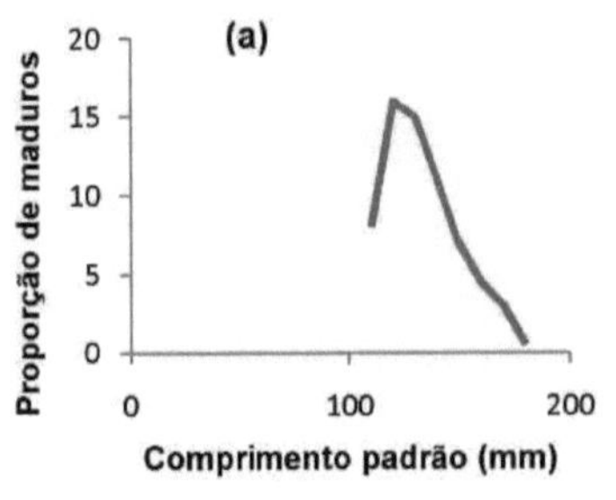


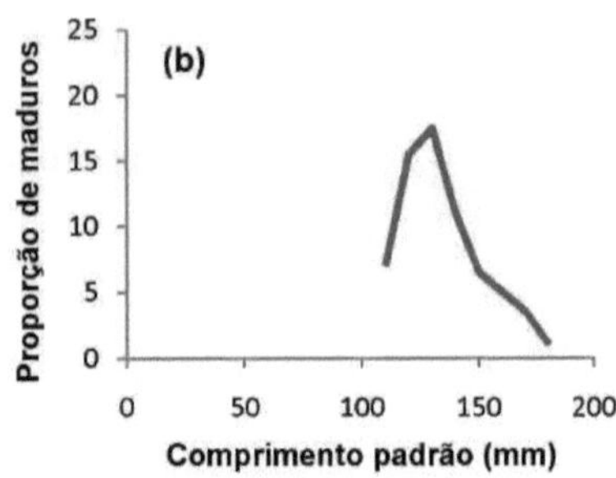


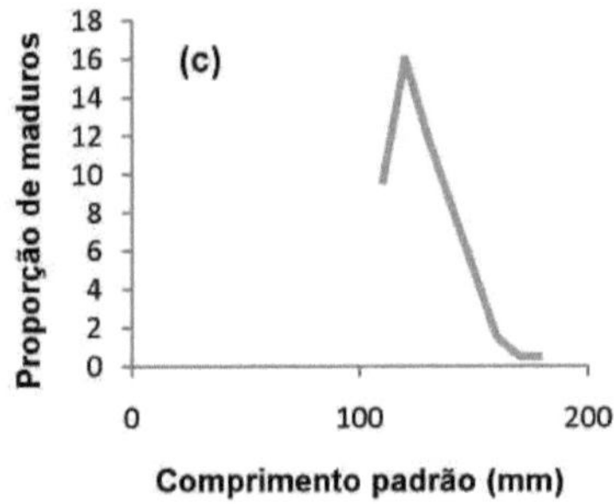


In terms of gonad morphology, *A. britskii* showed brown fusiform ovaries and whitish filiform testicles (Figure 10). During the study period, specimens were observed at all stages of maturation. In all the months of collection, the three reservoirs had immature (A), maturing (B) and mature (C) individuals. Stage D (spawned) was only observed in the PA I, II and III reservoirs in July 2015 and March 2016. The maturation of the gonads in all the months may be related to piecemeal spawning, which has already been observed as a strategy for the species (BARROS; SANTOS, 1996).

**Figure 11. Mature gonads of *Acestrorhynchus britskii*: ovaries (left) and testicles (right).**

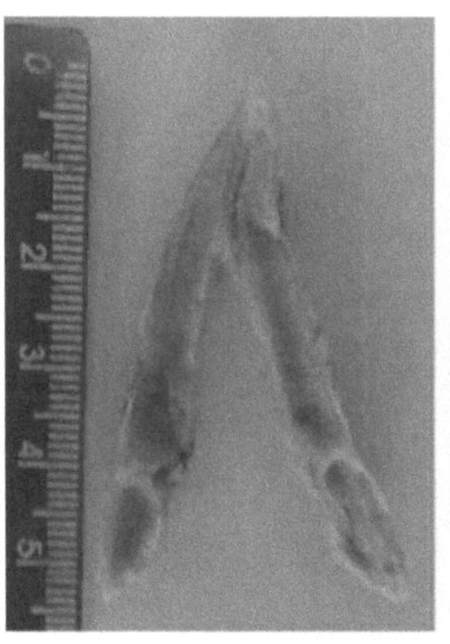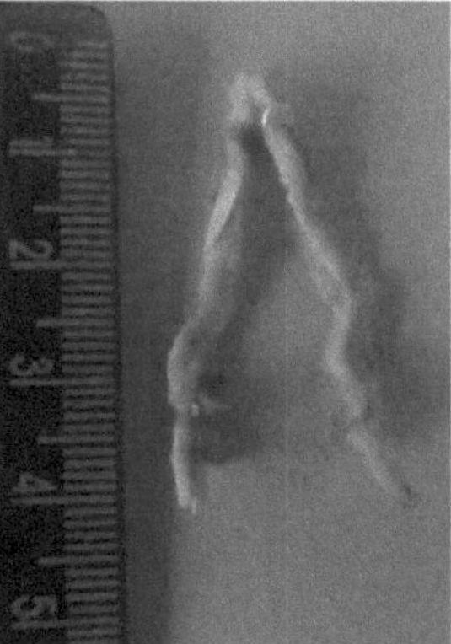

**Photos: SANTOS, 2016.**

In the Sobradinho reservoir the GSI varied between 1.5 (May/2016) and 5.4 (January/2016) with an average of 3.4 ± 1.4, in Itaparica it varied between 2.9 (July/2015) and 5.1 (January/2016) with an average of 3.8 ± 0.8 and in PA I, II and III it varied between 3.1 (May/2017) and 7.0 (January/2016) with an average of 4.4 ± 1.6. The t-test showed that there were no significant differences between the IGS averages in the three reservoirs (t=0.645536, p>0.05 between Sobradinho and Itaparica, t=0.328759, p>0.05 between Sobradinho and PA I, II and III and t= 0.470628, p>0.05, between Itaparica and PA I, II and III (Figure 12).

The highest GSI values were observed in January and March 2016, during the flood season. Soares *et al.* (2015) when analyzing the IGS of *Acestrorhynchus pantaneiro* in the Northern Pantanal (MT) also observed higher indices during the flood season, which indicates that the species of this genus may invest more in reproduction during this period. The averages revealed that in the PA I, II and III reservoirs the species showed greater reproductive investment, which may be related to the greater availability of food in the reservoir, a fact already observed by Santos, Gomes and Lopes (2006).

**Figure 12. Average IGS values for mature females of *Acestrorhynchus britskii* with standard deviation in the reservoirs of Sobradinho, Itaparica and PA I, II and III, Rio Sao Francisco, Brazil.**

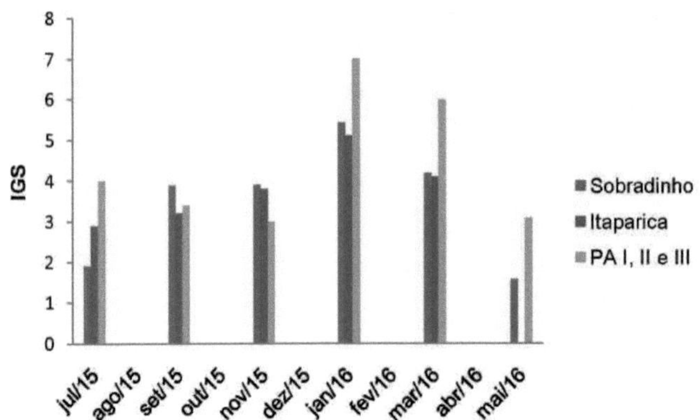


IGS
8
7
6
5
4
3
2
1
0
jul/15
ago/15
set/15
out/15
nov/15
dez/15
jan/16
fev/16
mar/16
abr/16
mai/16
Sobradinho
Itaparica
PA I, II e III

# 6  FINAL CONSIDERATIONS

According to the results, there were no significant differences between the three populations analyzed, suggesting that the survival conditions for the species are similar in the respective reservoirs, with the availability of resources and ideal reproduction conditions for *A. britskii to* continue maintaining viable populations.

The fact that it is not a migratory species lessens the impact of the dam on *A. britskii,* however, it should be noted that the new ecosystem formed starts to have a new dynamic depending on the type of operation of the dam, and the entire biota is affected. Anthropogenic activities such as the dumping of waste, industrial activities and the growing practice of aquaculture in the region where the study was carried out, together with irregular fishing, are also harmful factors that can affect the entire trophic chain of the environment.

The trophic position of *A. britskii* gives it the status of a fundamental component of the environment, forming part of the diet of larger piscivores and controlling populations of forage species. The study showed that the species is in a stable situation in the reservoirs studied, indicating that the impacts have not yet reached it in a considerable way by means of the parameters studied. Some factors observed, such as the maturation of the gonads in all the months of collection, in addition to being allied to its piecemeal spawning, show the species' intense reproductive activity, which may be a type of response to the changes found in the environment after the dam.

# 7    REFERENCES

ABNT - Brazilian Association of Technical Standards. **NBR 14724. informaçao e documentaçao: trabalhos acadêmicos: apresentação**. Rio de Janeiro, 2011.

AGOSTINHO, A. A., *et al.* **Ecologia e Manejo de Recursos Pesqueiros em Reservatórios do Brasil**. Maringà: Eduem. 2007.

AGOSTINHO, A. A.; PELICICE, F. M.; GOMES, L. C. **Dams and the fish fauna of the Neotropical region: impacts and management related to diversity and fisheries**. Brazilian Journal Biology. 2008.

ALVARENGA, E . R. *et al.* **Reproductive biology and feeding of *Curimatella lepidura* (Eigenmann & Eigenmann) (Pisces, Curimatidae) in Juramento Reservoir, Minas Gerais, Brazil. -** Revista Brasileira de Zoologia 23(2). 2006.

ALVES, C. B. M.; POMPEU, P. S. **Historical Changes in the Rio das Velhas Fish Fauna-Brazil**. American Fisheries Society Symposium. *In: The American Fisheries Society*, 2005. v. 45. 2005.

ANA - National Water Agency. **Sao Francisco Hydrographic Region**. 2016. Available at: http://www2.ana.gov.br/Paginas/portais/bacias/SaoFrancisco.aspx. Accessed on: 5 Jan 2017.

ARANTES, F. P., *et al.* **Profiles of sex steroids, fecundity, and spawning of the curimata-pacu *Prochilodus argenteus* in the Sao Francisco River, downstream from the Tres Marias Dam, Southeastern Brazil**. Animal Reproduction Science 118. 2010.

BARBIERI, G.; HARTZ, S.; VERANI, J. R. **The condition factor and hapatosomatic index as indicators of the spawning period of *Astyanax fasciatus* Cuvier, 1819, from the Lobo Reservoir, Sao Paulo (Osteichthyes: Characidae)**. Iheringia, Ser. Zoolo. Porto Alegre. 1996.

BARBOSA, J. M. **Faunal similarity between the Moji-Guaçu, Sao Francisco and Parnaiba river basins**. In: V Sem. Reg. de Ecologia, 1986, Sao Carlos, SP.

Proceedings of the V Sem. Reg. of Ecology. Sao Carlos, SP: UFCar, 1986.

BARBOSA, J. M.; SOARES, E. C. **Perfil da Ictiofauna da Bacia do Sao Francisco: Estudo Preliminar**. Rev. Bras. Enga. Pesca 4(1). 2009.

BARBOSA, A. S. **The environmental impacts of artificial reservoirs**. 2010.

Available at:

http://www.comitesm.sp.gov.br/erapido/arquivos/midia/1ef2af2a0d2f99297222abf65 9 b9373f.pdf . Accessed on; 04/04/2017.

BARROS, L. N. V.; SANTOS, G. B. **Fecundity and spawning aspects of dogfish Acestrorhynchus britskii Menezes, 1969 (Teleostei: Characidae)**. Arquivo Brasileiro de Medicina Veterinària e Zootecnia. 1996.

BEGON, M.; HARPER, J.L.; TOWSEND, C. R. **Ecology of individuals and ecosystems**. Artmed. 2007.

BENEDITO-CECILIO, E.; AGOSTINHO, A. A. **Structure of fish populations in the Segredo Reservoir**. In: AGOSTINHO, A. A.; GOMES, L. C. (Ed.).

Segredo Reservoir: ecological bases for management. 1 ed. Maringà: Eduem, 1997.

BERRA, T. M. **Freshwater fish distribution**. The Chicago University Press. 2007.

BRITSKI, H. A.; SATO, Y. ; ROSA, A. B. S. **Manual de Identificaçâo de Peixes da Regiao de Très Marias: com chaves de identificaçâo para os peixes da bacia do Sao Francisco.** 3rd Ed. Brasilia: Câmara dos deputados/CODEVASF. 19 88.

CARVALHO, P. A., *et al.* **Reproductive Biology of *Astyanax fasciatus* (Pisces: Characiformes) in a reservoir in southeastern Brazil**. - Journal of Applied Ichthyology 25. 2009.

CBHSF- São Francisco Hydrographic Basin Committee. **Regional Notebook**

**Hidrogràfica do Sao Francisco**. 2015. Available at:

http://www.mma.gov.br/estruturas/161/ publicacao/161 publicacao03032011023538 .pdf. Accessed on: January 31, 2017.

CHAVES, F. C., *et al.* **Reproductive dynamics and population structure of *Hoplias aff. Malabaricus* (Bloch, 1794)(Characiformes: Erythrinidae) in a reservoir of the Taperoà river basin, Paraiba**. Revista Biotemas. 2009.

CHESF - Companhia Hidrelétrica do Sao Francisco. **Chesf Sobradinho System**. 2017a. Available at:

http://www.chesf.gov.br/SistemaChesf/Pages/SistemaGeracao/Sobradinho.aspx. Accessed on: March 15, 2017.

. **Chesf Luiz Gonzaga System**. 2017b. Available at:

http://www.chesf.gov.br/SistemaChesf/Pages/SistemaGeracao/LuizGonzaga.aspx. Accessed on: March 15, 2017.

. **Chesf Paulo Afonso Complex System**. 2017c. Available at: http://www.chesf.gov.br/SistemaChesf/Pages/SistemaGeracao/ComplexoPauloAfo ns o.aspx. Accessed on: March 15, 2017.

CODEVASF - Companhia de Desenvolvimento dos Vales do Sao Francisco e do Parnaiba. **Inventory of irrigation projects**. 2ª ed. Brasilia: CODEVASF. 1991.

DE MÉRONA, B. ; ALBERT, P. **Ecological monitoring of fish assemblages downstream of a hydroelectric dam in French Guiana (South America).** Regulated Rivers: Research & Management 15, 1999.

ESCHMEYER, W. N.; FONG, J. D. **Species by Family/Subfamily**. 2016. Available at: http://researcharchive.calacademy.org/research/ichthyology/catalog/SpeciesByFam il y.asp. Accessed on: 27 Jan. 2017.

FADURPE - Apolônio Salles Foundation for Educational Development.

**Report on limnological monitoring and fishery production in the Sobradinho

**Reservoir**, Subproject: Fishery biology studies in the Sobradinho Reservoir. Final Ichthyofauna Report. 2004.

GODINHO, H. P., GODINHO, A. L. **Ecology and conservation of fish in southeastern Brazilian river basins submitted to hydroelectric impoundments**. Acta Limnologica Brasiliensia 5, 1994.

GODINHO, A. L.; GODINHO, H. P. **A brief view of the São Francisco**. In:

GODINHO, H. P.; GODINHO, A. L. (Org.) Aguas, peixes e Pescadores do Sao Francisco das Minas Gerais. Belo Horizonte: PUC Minas. 2003.

GODINHO, A.L. **Lista de peixes da bacia do Sao Francisco**. 2009. Available at: www.saofrancisco.bio.br. Accessed on April 28, 2017.

GONÇALVES, T. L., *et al.* **Gametogenesis and reproduction of the matrinxa *Brycon orthotaenia* (Günther, 1864) (Pisces: Characidae) in the Sao Francisco River, Minas Gerais, Brazil.** Brazilian Journal of Biology 66 (2A). 2006.

GOTELLI, Nicholas J.; ELLISON, Aaron M. **Principles of Statistics in Ecology**.

Translation: Fabricio Beggiato Baccaro et al. Porto Alegre: Artmed. 2011.

GUBIANI, E. A.; HORLANDO, S. S. **Length-weight and length-length relationships and length at first maturity for freshwater fish species of the Salto Santiago Reservoir, Iguaçu River Basin, Brazil**. Journal of Applied Ichthyology, v. 30, 2014.

GURGEL, H. C. B. **Population structure and breeding season of *Astyanax fasciatus* (Cuvier) (Characidae, Tetragonopterinae) from the Ceará-mirim River, Poço Branco, Rio Grande do Norte, Brazil**. Brazilian Journal of Zoology. 2004.

HASTINGS, P. A.; WALKER, H. J.; GALLAND, G. R. **Fishes: A guide to their diversity**. University of California Press. 2014.

HELFMAN, G. S. *et al.* **The diversity of fishes: Biology, Evolution and Ecology.**

Willey-Blackwell . 2nd ed. 2009.

HILSDORF, A.; PETRERE , M. **Conservaçao de peixes na bacia do Rio Paraiba do Sul**. Ciência Hoje Magazine, v.30. 2002.

HOJO, R. E. S., *et al.* **Reproductive biology of Moenkhausia intermedia (Eigenmann) (Pisces, Characiformes) in Itumbiara Reservoir, Goiàs, Brazil.** - Revista Brasileira de Zoologia 21(3). 2004.

LE CREN, E. D. **The length-weight relationship and seasonal cycle in gonad weight and condition in the perch Percafluviatilis**. J. Anim. Ecol. 1951.

LIMA, A. E. ; SEVERI, W. **Trophic state in the cascade of reservoirs of a river in the Brazilian semi-arid region**. Ver. Bras. Ciên. Agrâr. Recife. 2014.

LIZAMA, M. A. P.; AMBRÒSIO, M. A. **Weight-length relationship and population structure of nine species of Characidae in the Upper Paraná River floodplain, Brazil**. Brazilian Journal of Zoology. W99.

LOWE-MCCONNELL, R. **Ecological studies in tropical fish communities**. Sao Paulo: EDUSP, 1999.

LÜTKEN, C.F. **Velhas-Flondens Fisks**. Over. Dansk. Vid. Forh. Kjobenhav. v.12. 1875.

MELO, R. M. C. *et al.* **Comparative Morphology of Gonadal Structure Related to Reproductive Strategies in Six Siluriform Fish Species from the Sâo Francisco River Basin**. Brazil. Rev. Bras. de zool. 2011.

MENEZES, N. A. **Taxonomic redefinition of *Acestrorhynchus* species of the *lacustris* group with the description of a species (Osteichthyes, Characiformes, Characidae)**. Commun. Mus. Ciênc. PUCRS, Sér. Zool. Porto Alegre. 1992.

MENEZES, N. A. **Family Acestrorhynchidae**. 2003. In: REIS, R. E., KULLANDER, S. O.; FERRARIS, C. J., Check list of the Freshwater of South and Central America. Porto Alegre: Edipucrs. 2003.

MONTENEGRO, A. K. A. **Population and feeding structure of *Steindachnerina notonota* Miranda Ribeiro 1937 (Actinopterygii, Characiformes, Curimatidae) in the Taperoâ II reservoir, in the semi-arid region of Paraiba, Brazil**. Acta Limnològica Brasiliensia. 2011.

MOTA, T. F. M.; PRIOLI, S. M. A. P.; PRIOLI, A. J. **Phylogenetic studies of the order Characiformes: trends and shortcomings**. Publ. UEPG Ci. Biol. Saùde, Ponta Grossa, v.20, n.1. 2014.

NASCIMENTO, R. S. S. ; GURGEL, H. C. B. **Population structure of Poecillia vivipara Bloch; Schneider, 1801 (Atheriniformes, Poecilidae) from the Ceará-Mirim River - Rio Grande do Norte**. Acta Scientiarum ; Biological Sciences. 2000.

NASCIMENTO, W. S., YAMAMOTO, M. E.; CHELLAPPA, S. **Sex Ratio and**

**Weight-Length Relationship of the Annual Fish *Hypsolebias antenori* (Cyprinodontiformes: Rivulidae) from Temporary Pools in the Semi-Arid Region of Brazil**. Biota Amazônia. V.2, n.1.2012.

NELSON, J. S. **Fishes of the world**. Jhon Willey & Sons, 4ª ed. 2006.

NIKOLSKII, G. V. **Theory of fish population dynamics: as the biological background for rational exploitation and management of fishery resources**. Edinburgh: Oliver & Boyd, 1969.

ODUM, E. P. **Fundamentals of Ecology**. 6th Ed. Guanabara Koogan, Rio de Janeiro. 2004.

ORSI, M. L; SHIBATTA, O. A.; SILVA-SOUZA, A. T. **Biological characterization of fish populations in the Tibagi River, Sertanópolis locality,** in :MEDRI, M.

E. The Tibagi river basin, Londrina, State University of Londrina. 2002.

PAIVA, M. P. **Relationship between the number of predatory fish species and 24 fish yield in large northeastern Brazilian reservoirs**. In COWX, IG. (Org.).

Rehabilitation of freshwater fisheries. Osney Mead:Fishing News Books. 1994.

PERET, A. M. **Feeding dynamics of piscivorous fish in the Très Marias reservoir - MG**. Dissertation (Master's Degree in Ecology and Natural Resources)

Federal University of Sao Carlos-SP. 2004.

PETRERE, M. **Fisheries in large tropical reservoirs in South America**. Lakes & Reservoirs: Research & Management 2. 1996.

PRETTI, V. Q. *et al.* **Phylogeny of the Neotropical genus *Acestrorhynchus* (Ostariophysi: Characiformes) based on nuclear and mitochondrial genes sequences and morphology: A total evidence approach.** Molecular Phylogenetics and Evolution, V. 52, 2009.

RAPOSO, R. M. G.; GURGEL, H. C. B. **Population structure of *Serrasalmus spilopleura* Kner, 1860 (Pisces: Serrasalmidae) from the Extremoz lagoon, Rio Grande do Norte, Brazil**. Acta Scientiarum, Maringâ. 2001.

REIS, R. E. ; KULLANDER, S. O. ; FERRARIS, C. J. **Check list of the freshwater fishes of South and Central America**. Porto Alegre: Edipucrs. 2003.

RICKLEFS, R. E. **The Economy of Nature**. Guanabara Koogan. Rio de Janeiro. 2010.

ROCHA, A. A. F., *et al.* **Trophic relationships between *Acestrorhynchus britskii* (native) and *Plagioscion squamosissimus* (introduced) in cascading reservoir systems**. Bol. Inst. Pesca, Sao Paulo. 2015.

SANTOS, A. C. A.; NOVAES, J. L. C. **Population structure of two *Astyanax* Baird; Girard, 1854 (Teleostei: Characidae) species from upper Paraguaçu river.**

Tecpar. 2008.

SANTOS, T. S; VILHENA-PICANÇO, M. D. ; SA-OLIVEIRA, J. C. 2007. **Weight-length relationship of *Acestrorhynchus lacustris*, Lütkèn, 1875 (Characiformes: Acestrorhynchidae) from the Curiaù River APA, Macapà- AP**. Proceedings of the VIII Congress of Ecology of Brazil. Caxambu-MG. 2007.

SANTOS, E.; GOMES, S. O.; LOPES, J. P. **Contribution of the eelgrass *Egeria densa* to fish farming through the colonization of the shrimp *Macrobrachium amazonicum* in the sub-medium São Francisco River, Northeastern Brazil.** Rev. Bras. Eng, Pesca. 2006.

SARMIENTO, F. O. **Diccionàrio de Ecologia: paisajes, conservàcion e desarrollo sustentable para Latinoamérica.** Editorial Abya Yala. 2000.

SATO, Y. ; GODINHO, H. P.. **Fish from the São Francisco River basin.** In: LOWE- McCONNELL, R. H.. Ecological studies of tropical fish communities. Sao Paulo: Edusp, 1999.

SATO, Y. (2005) **Influence of the Abaeté River on the reproductive success of the neotropical migratory teleost Prochilodus argenteus in the Sao Francisco River, downstream from the Três Marias Dam, southeastern Brazil.** River Research and Applications, V. 21. 2005.

SILVA, J. P. A., *et al.* **Reproductive tactics used by the Lambari *Astyanax aff. fasciatusin* three water supply reservoirs in the same geographic region of the upper Iguaçu River.** Neotropical Ichthyology, v.8, n. 4. 2010.

SOARES, K. M. *et al.* **The flood pulse in the Caiçara bay system, Northern Pantanal Mato-Grossense (Brazil), determines the reproduction of *Acestrorhynchus pantaneiro* (Menezes, 1992) (Characiformes, Acestrorhynchidae).** Rev. Acad. Ciên.Anim. 2015.

SOUZA, R. A. L.J. *et al.***Development of tambaqui (*Colossoma macropomum* Cuvier) (Pisces, Characidae) raised in the Guamà River, State of Parà, Brazil.** CEPTA Technical Bulletin. 2000.

SUZUKI, H. I. *et al.* **Inter-annual variations in the abundance of young-of-the-year of migratory fishes in the Upper Paranà River floodplain: relations with hydrographic attributes.** Brazilian Journal of Biology, v. 69 (2, Suppl.), p.649-660, 2009.

THOMÉ, R. G., *et al.* **Reproductive biology of *Leporinus taeniatus* Lütken (Pisces, Anostomidae) in Juramento Reservoir, Sao Francisco River basin, Minas Gerais, Southeastern Brazil** - Revista Brasileira de Zoologia 22(3). 2005.

TOS, C. D. *et al.* **Variation of the ichthyofauna along the Goioerê River: an important tributary of the Piquiri-Paranà basin**. Iheringia: Série Zoologia, v. 104, n. 1, 2014.

TRAVASSOS, H. **Catàlogo dois peixes do vale do rio Sao Francisco**. Bol. Soc. Cear. Agron.1960.

VAZZOLER, A. E. A. M. **Biologia da reproduçao de peixes teleósteos: Teoria e Pràtica**. Maringà, Eduem. 1996.

VIEIRA, F. *et al.* **Fish from the Iron Quadrangle: Identification Guide.** Biodiversitas Foundation, Belo Horizonte-MG. 2015.

WOOTTON, R. J. **Ecology of teleost fishes**. Chapman and Hall (London). 1990.

Printed by Books on Demand GmbH, Norderstedt / Germany